Hammad's Theory On Spirit and Soul

Elsebai Hammad

Published by Elsebai Hammad, 2025.

HAMMAD'S THEORY ON SPIRIT AND SOUL

First edition. November 30, 2025.

ISBN: 979-8233840647

Written by Elsebai Hammad.

Introduction

This booklet addresses one of the most important and touching subjects that concern all inhabitants of earth, i.e. the "Rouh", which is responsible for life in humans and animals. We tried to discuss it in a logical, scientific, and trans-religious way. Although we intended to keep this work purely scientific, some religious and folk material inevitably came across.

Thousands of scholars have tackled this subject before and left infinite number of thoughts that were mostly theoretical and vague, not even one was convincing to the author because they were not exposed to modern sciences, and most of them were not physicians.

The word "Rouh" is used in the Arab and Muslim world to express something mysterious related to life and death, and most of their dictionaries define the Rouh as one thing and one entity. However, in western countries and most other parts of the world, two terms are commonly used instead: "spirit" and "soul". In our view, the "Rouh" is a collective term containing three different entities: spirit, soul and secret of life, which is a new entity we have added for the first time in this theory, and each one of these components is different.

We hope this new theory will set the records straight, make terms reasonably clear, and is appreciated by readers.

Note: An Arabic edition of this theory was issued in 2022 by Dar Alhelal Publishing Institution in Cairo.

Elsebai F. Hammad, MD, PhD

Cairo, Oct. 20, 2025

Elsebai_Hammad@yahoo.com

Historical note

Throughout history, philosophers, clergy and laymen have struggled to understand the nature of spirit and/or soul. The Pharaohs, for example, had complex ideas about this vague topic and divided it into several components, one of them as a bird with a human head, and believed in its return to the body and in resurrection, recording that in their "Book .of the Dead" and on walls of tombs and temples

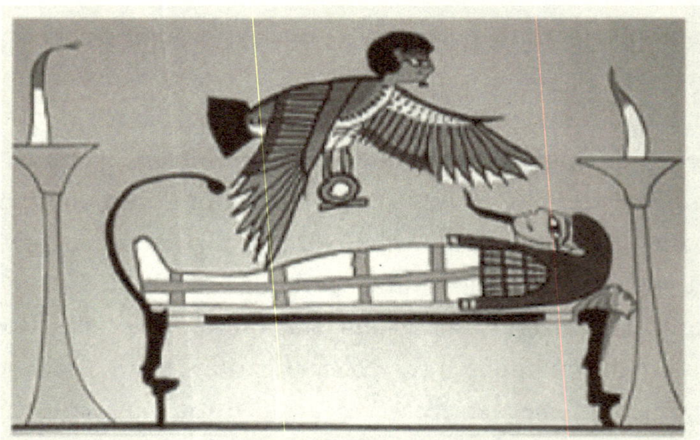

Figure (1) The Soul is represented as a bird with a human head in the ancient Egyptian civilization

After the pharaohs, many other civilizations and religions have emerged such as the Greek, the Roman, the Hindu, the Muslim, the Christian, and the Jewish, all of them have added thoughts to the subject but were .mostly theoretical, and there is no room to review all of them here

Throughout the years, two major obstacles have faced this complex topic, the vague terminology and viewing it as a spiritual not a scientific issue.

In the early Islamic era, some scholars have added more reflections, but still there was disagreement between them about the interpretation of the verse in the Quran (Esraa 85): "and they ask you about the Rouh, say the Rouh is from the command of my Lord, and you have not been given knowledge except a little".

A group of Muslim scholars, including Al-Sahrawardi, said that since the Rouh was something from the Lord's command, it could not be spoken of, while some others said that the text admits that human beings had been given little knowledge, so why don't we speak within the limits of this little knowledge, and since the term "Rouh" has more than one meaning and use, among those was the Faqeeh Ibn Qayyim Al-Jawzyyia who wrote a book in about 350 pages seven centuries ago titled "AlRouh", but this book was met with severe criticism because it relied on dreams, tales, talks and even poetry, also his book was issued at a time when medicine was primitive "folk medicine" relying on herbs, amulets, spiritual practices and moxibustion, therefore lacking any true medical information.

Figure (2) Examples of Old and recent Books on spirit and soul.

(Ref 5, 6)

Sadly, since the days of the ancient Islamic scholars up till now, there has been disagreement over the meaning of the words "soul" and "spirit". Most of the Arabic dictionaries are still confused and say that the two words have the same meaning (Rouh). This is not reasonable or acceptable, because people who go into and out of coma, staying for months or years in deep unconsciousness prove the difference. In coma the spirit exists but soul is paralyzed. However, a few Arab philosophers were aware of that difference, saying that the two words are different, where "the spirit is light from God, while the soul (or self) is clay" a view that seems to be scientifically correct and accepted in many other world cultures. Whether "mind" is a part of Rouh or not is controversial, but we think that mind is a tool or a program in the brain dedicated to serving the components of Rouh.

Accordingly, the Arabic word "AlRouh" should appropriately have at least two components: a divine or secret part that controls the moment of death, known as "spirit," and another sensible "personal" part called "soul" with all its lusts and diseases. These two parts or components lie within a mortal body and together constitute what is known as "the body entity".

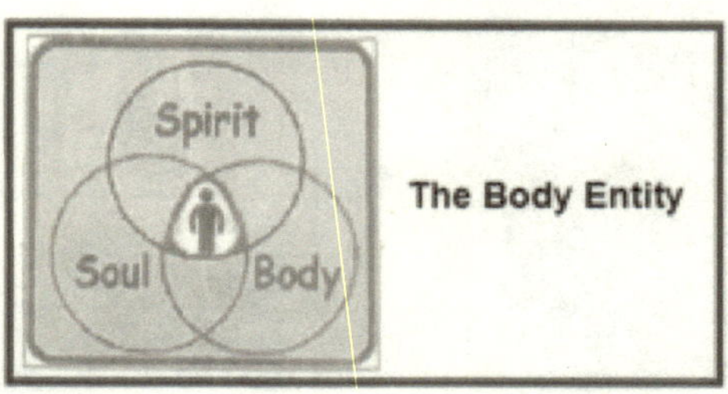

Figure (3) The three components of the body entity

Admitting that "Rouh" has two parts, spirit and soul, as mentioned above, is consistent with modern thinking and resolves the debate over the definition in the Arabic dictionaries, and pave the way for deeper research in the subject.

As author, having long experience in teaching anatomy and embryology, dealing with cadavers for decades and doing research and writing books, especially the one titled "The Rouh and Modern Medicine" (2015), I became convinced that a "theory" is necessary for such a difficult and complex subject. And surely, a physician's view here would be different from that of non-physicians.

Basics of the theory

Following each death, amputation, or transfer of one organ from one body to another, a few questions arise: where are the spirit and soul, do they have specific location in the body, and if so, where we find them in conjoined twins for instance?

Initially, let's think of spirit and soul separately. The spirit is the mysterious part related mostly to the moment of death, and known in the colloquial language as the "God's Secret", but where is it found and how it works in the human body? To trace what the spirit is, we go back to the book of Ibn Qayyim. Three lines in this book drew the author's attention because they contained a helpful scientific idea about the relationship between Rouh and body when he said "the Rouh either covers all the body from outside like a dress, or spreads into the entire body, or resides into one place such as the heart or the brain", although he was inclined to the second possibility (spreading into the whole body).

In the light of recent developments in modern medicine, we can revise those three possibilities and exclude the unreasonable ones, to reach the most reliable one.

Amputations and organ transplantations that take place in hospitals on a daily basis, in which any part of the body can be removed, do not go along with a spirit that covers the whole body from outside or spreads into the entire body either, because cutting or removing parts of body this way means cutting or removing parts of spirit which seems illogical and impossible because it must be an indivisible entity. So, the one-site hypothesis of spirit location remains significant, i.e. in the heart or in the brain.

First let us think of a spirit located in the heart. This also seems impossible too after the techniques of heart transplantation and the possibility of a man living without a heart for 35 days, connected to an artificial heart machine as reported in this case (https:\\hight3ch.com>the-first-man-who-lived-without-a-heart)

Thus, a spirit located in the heart is again excluded.

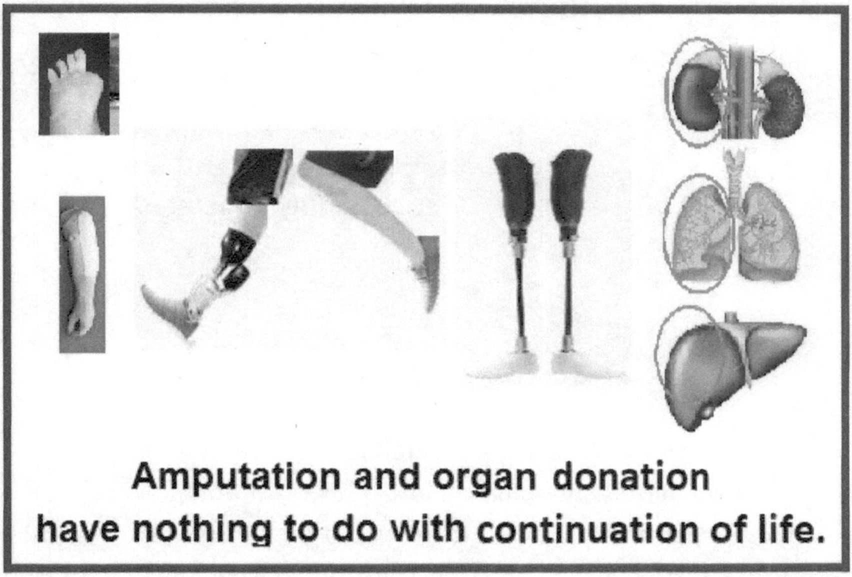

Figure (4) Different body parts that can be amputated or replaced with transplant

Finally, the last and only remaining hypothesis is a spirit located in the brain. To confirm and validate this hypothesis, we have carried out an in-depth study of conjoined twins attempting to understand the number of spirits in the various forms and degrees of body adhesion. Our research in this area has yielded very fruitful results.

The two European girls connected to each other through the Pelvic region were successfully separated and each of them made her way through life. They provided soundproof that each of them had her own spirit and is located in her head.

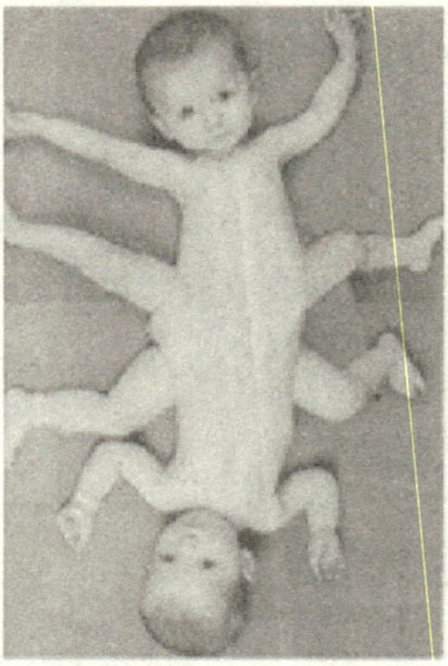

Two girls were connected at the pelvic region and were successfuly separated.

Figure (5-A) Conjoined twins

But when the head is doubled on one body or in double lower half of the body under a single head, there would be considerable confusion and require further analysis and investigation. The conjoined twin Abby and Brittany Hensel are a good example. Here we see two heads on one body, each head has a name, and from the news and research published on their case, it turned out that each head is different in thinking, ideas, and inclinations. One head is skilled in linguistics, and the other is more efficient in mathematics, they differ also in the sense of humor, although the body is one with one genital system. This gives clear evidence with no doubt that there is one different spirit in each head.

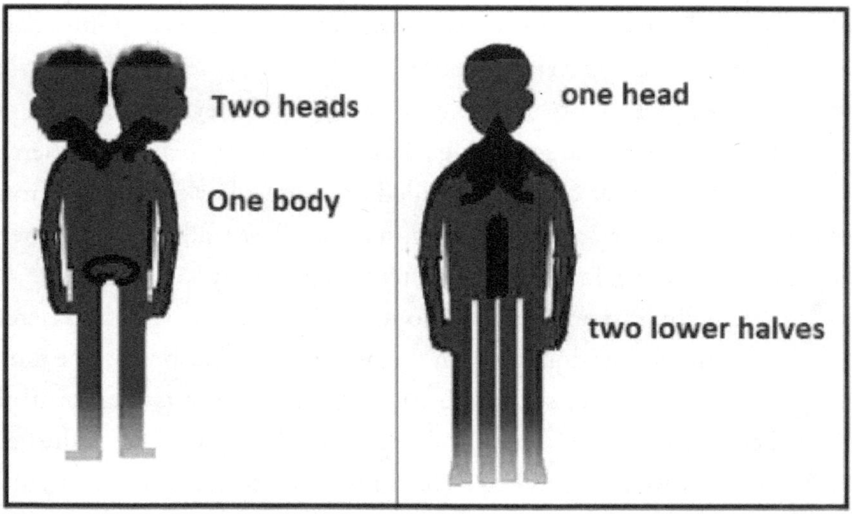

Figure (5-B) Conjoined Twins

Lady Myrtle Corbin is another fascinating example. This beautiful American woman who lived in the early part of the last century was born with tow pelvis, four legs and one head. She was married to a doctor and gave birth to five children, three from the right vagina and two from the left one. Her brain, feelings, and thinking, all come from one head with one spirit in it.

As we have seen, all kinds of conjoined twins support the last hypothesis, and confirm that the secret part or the "God's secret" is within, and works through the head, but another question arises: in which part of the head? This led the author to propose the "Exclusion Theory," as a first step to understanding what spirit and soul are, and how they behave in the human body. In other words, to localize the smallest area or region in the body that is convenient for spirit to perform its duties and use it as "command center".

The method of exclusion is based on medical facts, since it has been proved that more than 99% of the body parts are service-oriented or service-based. The four limbs for example, all might be amputated following certain accidents, or could be totally absent in some forms of congenital anomalies (Amelia), meaning that the four limbs have nothing to do with the continuation of life. The same for most of the intestine, and most of organs like lung and liver, kidney, ears, eyes etc. all could be removed without causing death, until we reach a point where a very small region that can't be excluded or removed or transferred from one body to another, and its destruction means death, all these characters apply to a part of brain called "brainstem" or part of it.

Since the brainstem contains too many vital centers and considered to be "government headquarters" in the human body, harboring the most dangerous vital centers such as breathing and cardiovascular control, sleep, etc., we can again apply the exclusion theory on the brainstem itself. This way we have reached a conclusion that the brainstem reticular formation (RF) or one of its large nuclei which gives more orders than it receives, unlike the rest of the centers that exchange information by taking and responding in a semi-balanced manner, is intended.

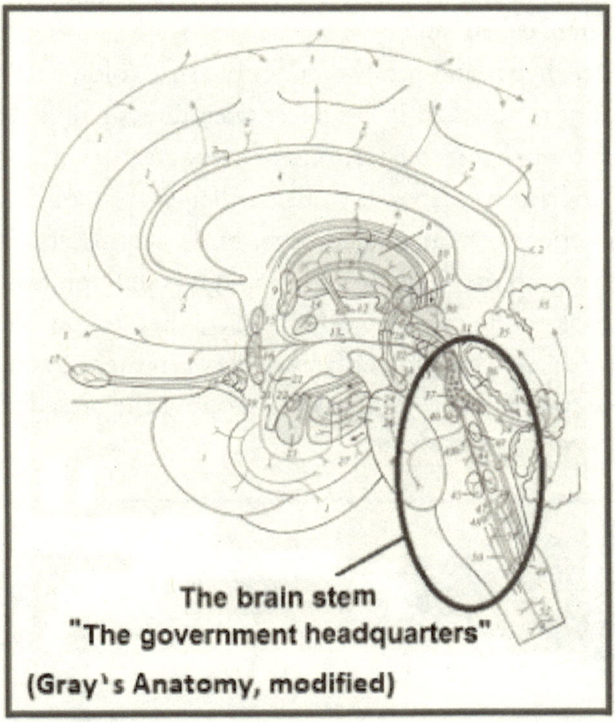

Figure (6) The brainstem (Ref 1)

This explains why sudden death occurs in breaking the neck (in death by hanging), and why the brainstem is the last to become silent in the body, that is then the person is pronounced dead.

Now, after locating the most sensitive and crucial command center in brainstem from which we are convinced that the first part of Rouh or the spirit works, we come to the second part of Rouh in our theory: The soul, what it is? And how does it work?

Here, and perhaps for the first time, we concluded that the second part of Rouh which is the soul, is nothing but the DNA profile which is the active, essential and integral part of one's life and is expressed from birth to death as the "self". The DNA helix is the basis and the constitution of life for all living creatures no matter how small or big they are, and it has great interaction potential with the environment via a brain program called "mid".

The DNA helix (and print) dictates the size, color, behavior and all manifestations of life, and it has never been directly linked to the soul before, even though it is the portfolio of every creature on earth.

To generalize our theory, it would be helpful to look at some living creatures inferior to humans like dogs, birds, worms etc. Surely these creatures have spirits and souls although some philosophers and physicists look at them differently. In fact, they follow the same life dogma as we do, and the only difference between them and humans, and between themselves is the number of neurons or the presence or absence of brain.

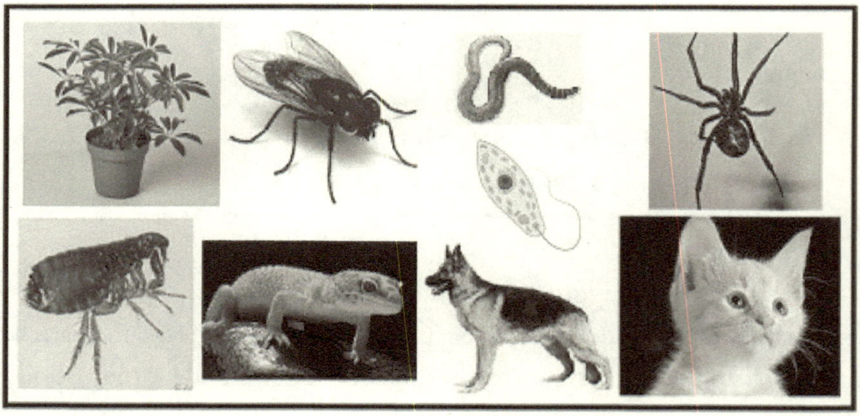

Figure (7) Different living creatures

Those animals who have heads, have spirits working from their heads, and a DNA helix (genes) expressing their identity or self, and performing reproduction as well. The full record of the whole organism's shape, character and behavior regardless of its size, is recorded and found in his fertilized egg (combined male and female DNA), others who depend on simpler ways of reproduction i.e. asexual, still use the DNA helix for the same purpose because it is the only tool of life.

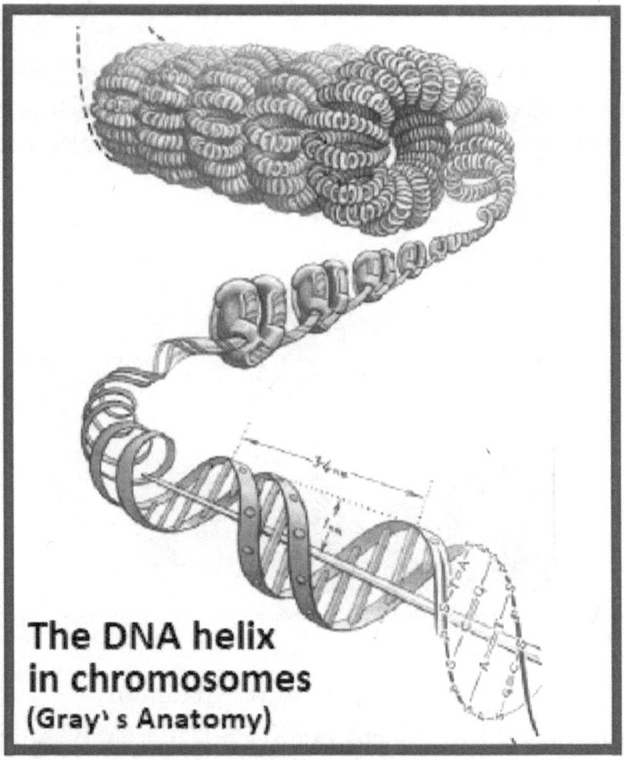

Figure (8) The DNA Helix

Figure (8) depicts the DNA helix that contains all the properties of the organism: color, size, gait, sound, and all the rest of traits recorded on a huge number of genes. This recorded program is the milestone that is formed directly because of fertilization of an egg with a sperm and is implemented in the womb just as a printer taking orders from a computer. Small creatures like bacteria, viruses, fungi, protozoa, and even sperms, all lack a central nervous system or neurons but can sense, react, and communicate with each other, directed by their sophisticated DNA sequencing. The size of information in the DNA molecule is about 6.2 billion bites.

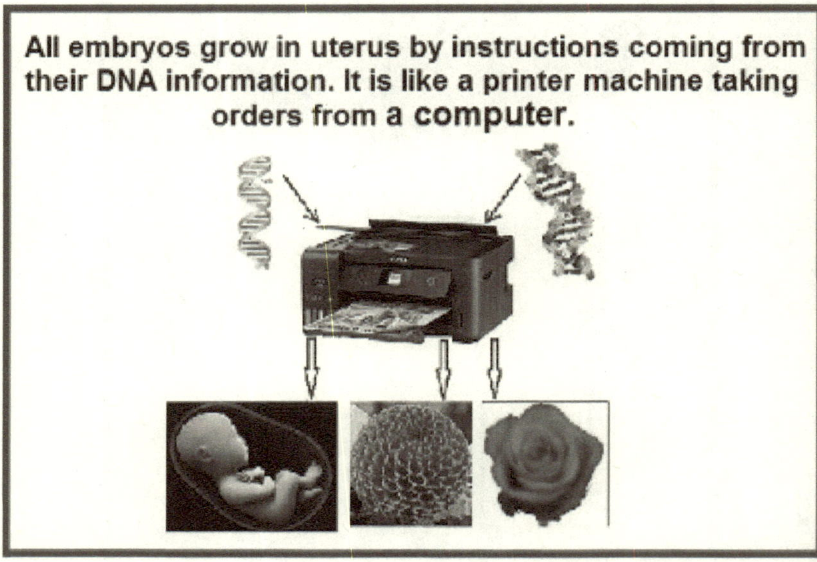

Figer (9) DNA printing function

The length of DNA implementation in uterus is linked and is proportionally related to the size of the animal, Figure (10).

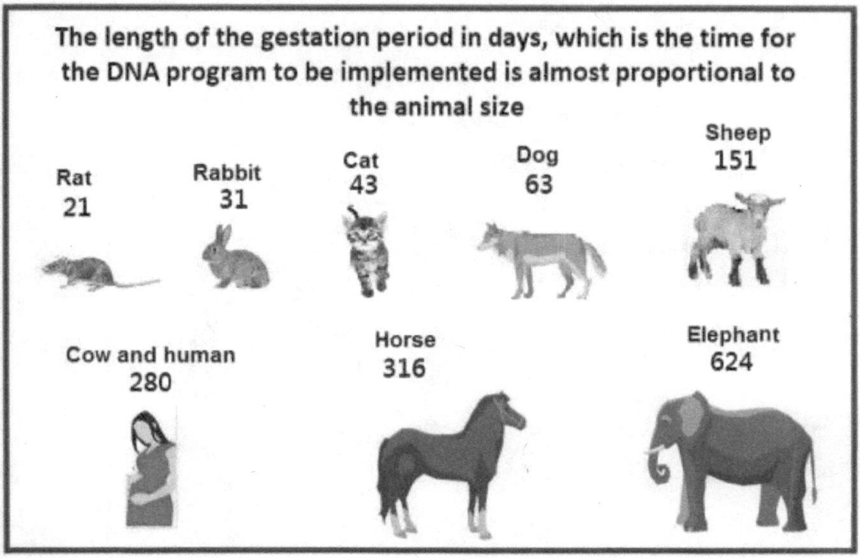

Figure (10) DNA implementation period in days (pregnancy duration)

All living organisms share parts of the DNA material. We humans differ from each other in size, colors, tongues and traits, all such variations are attributed to a tiny proportion of our DNA (its cellular products and the number of active genes). The variation gradually widens as we go down through the animal kingdom. Between man and chimpanzee, the difference is about 1%, for rats 15%, for cows 20%, and even flies and bananas share about 60% of our DNA material.

Figure (11) Variable products of DNA material after egg fertilization

Also, it is the DNA fingerprint pattern that gives flowers, butterflies, fish and shellfish their pretty colors, and to animals their shapes, sizes, behaviors, even blood groups, also the type of food and the design of beak in birds, all are determined by the DNA sequence of genes.

Figure (12) Types of birds are affected by DNA pattern

And, as we mentioned earlier, animals that are too small and lack heads, as well as plants, their spirits and souls are solely in their DNA patterns.

All living things (organisms) are completely governed by and succumbed to their DNA orders coming from within, and in the meantime are placed in an environment that could be against their will with rules and laws which must be observed and abided by. Such a fine process of balancing is the mystery and the test of the whole life.

Acknowledging, with respect, the role of religions in this aspect, many of them have tackled the issue of Rouh (sprit and soul) from various perspective and we have examined some of their views and found that a few of them were close to modern science and close to our theory, others were not. In Islam, for example, the DNA had been introduced into the subject of the spirit and soul, although indirectly, through the prophet Muhammad's saying: "Every son of Adam is eaten by earth except the tailbone from which he was created, and from which will be resurrected" (a prophet Muhammad's saying). The tailbone is the hardest piece in skeleton and like many other body remnants contain DNA traces. Surprisingly, the human embryo grows and spreads from this location via its main middle layer (mesoderm) superiorly and laterally like a tree in the sixteenth day in uterus.

Again, it is worth mentioning that the growth of an embryo is controlled solely by its DNA, and it is now possible to extract DNA samples from ancient remains thousands of years old to be read, decoded and translated.

Growth of the main middle layer of embryo (mesoderm) lika a tree from the primitive node

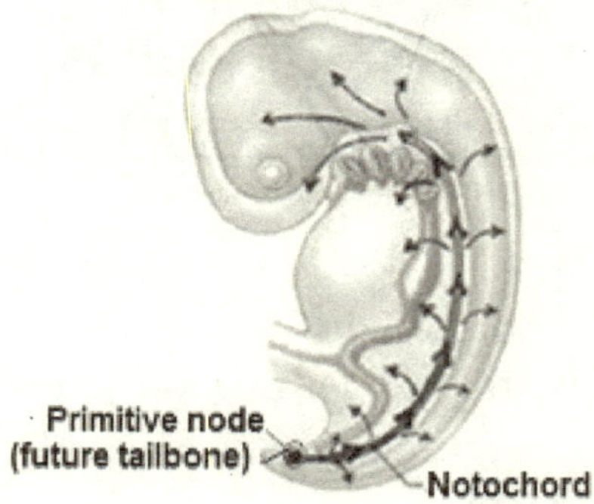

(Medical Embryology, modified)

Figure (13) Human embryo (Ref 2)

The human soul is basically DNA product and is amalgamated with brainstem performance where the spirit is located. And, if we ask why the embryo grows in the womb like a tree from this point, and why would he grow again one day later from the same point mentioned above? The answer and the secret lie in the DNA molecule.

It has become clear now that resurrection is not only a religious or imaginary process, but it is now a fact after cloning of animals and possibly men using just one molecule of their DNA in the process. Also, the DNA molecule seems to have supernatural forces in addition to its functions that we know, especially the subconscious and unconscious minds which we know little about. Yoga for example has been around for thousands of years, they say it creates a balance between the mind, body, and soul. Also, hypnotherapy and placebo are effective methods of treatment but not fully understood. Moreover, some researchers have suggested that energy waves and spectra possibly emitted from DNA material of those who died and left them scattered all over the ground during their lives, are behind visits and voices of the dead that we see and hear every now and then, something relating or similar to halos and colors proved to be emanated from an individual's 'energy field' during life.

DNA is becoming a perfect tool for identification as well, and who knows, it may become mandatory to present the DNA print as an identification card at the gates of sky and heaven later! (as a "Cosmic" ID).

Although the DNA composition is unique to every living organism, it gives up some of its privacy during organ transplantation from one person to another. Yes, it's the backbone of a person's spirit and soul, but when an organ is moved and placed into another body having different DNA in its cells, something amazing happens. This organ will adhere and do the same function in the new body although it came from different one. A property or instinct related to the donor's DNA, but it is another third and different property other than spirit and soul, we identified and named it here for the first time as "The secret of life". This secret helps organs to stick and continue to function in a new different body having different DNA, although it requires some tissue compatibility tests to avoid immune reaction and rejection in the recipient body.

To get a strongest "secret of life" in donated organs from living or dead donors, it is extremely necessary to remove and plant them as quickly as possible or to freeze them until surgery is convenient, this is because of the rapid autolysis and deterioration that takes place in tissues as time lapses after death. This secret of life totally disappears from organs after a week or 10 days, unless frozen as we mentioned above.

At the moment of death, the spirit "Gets out" in the colloquial terms, but we, physicians and scientists do not know exactly what comes out from the body at this moment, but we do know for sure the cause of death: that is organ failure followed by brainstem failure.

Back to some of the valuable religious views other than Islam about spirit and soul. In the Catholic Christian faith, an extremely accurate and scientifically correct doctrine says: "every human being receives his soul at the moment of conception", i.e. equally from father and mother. This is correct and is consistent with our theory and the role of DNA that we assumed and means that the fertilized egg contains a full record for the new person.

In the Book of Tawism, we find additional emphasis on the role and relationship between DNA and spirit and/or soul. According to their belief, spirit has three sources: "part from father, part from mother, and an original third part".

All of these "religious" views were said before sciences were developed, but some of them came close and hinted to a substance that was only discovered in 1953, called DNA (Deoxyribonucleic acid).

Animal and plant behaviors align with our theory

Animal behaviors like play, courtship rituals, parental care, territorial defense, aggression, migration, camouflage, and adaptation, all of them should not be taken for granted or haphazardly, because they all are "constitutional" or genetic i.e. set up by the law of the creator when the earth was created.

Small creatures that lack heads, brains or neurons like bacteria and other unicellular organisms as well as plants, perform elegant and highly sophisticated jobs that we humans cannot understand, perform, or even think of. They are often described as "living computers". Bacteria, for example, learn and have memory, sense a lack of food and reduce reproduction, know when to attack or migrate, and can adapt to harsh conditions through genetic mutation, developing different strategies for survival. All these actions require "thinking", "intelligence", and "brain", that are there, but we don't see them.

Neutrophils, one type of our white blood cells, is another example. They perform so many tasks; respond and move to the site of infection where they engulf bacteria, release enzymes and antimicrobial substances, and even help in the healing process by cleaning debris!

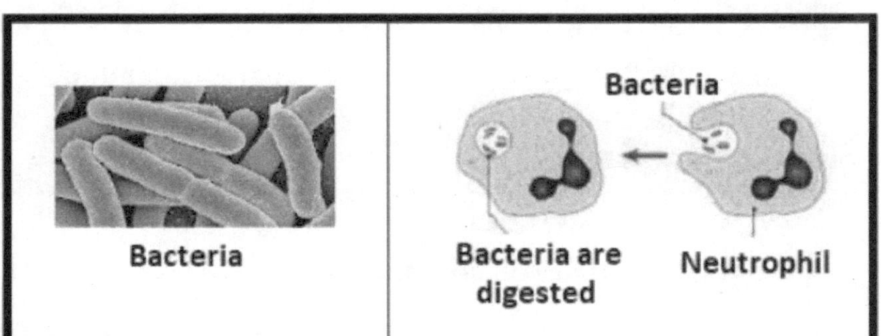

Figure (14) Bacteria and neutrophils

Additional fascinating example is found in the liver cell which contains tiny sacs called lysosomes that are filled with hydrolytic enzymes. Their function is to collect waste. They roam within the cell on their own and collect waste that constantly falls during the cell metabolism and the remnants of aging organelles, exactly like sanitation workers. Not only that, but they recycle the waste and make use of its components. Interestingly also, they multiply in certain diseases! Then, we have the right to ask: who is the "mayor" of the cell that declared a state of emergency and requested more new workers to deal with the catastrophe? There must be a kind of mind there. The way lysosomes work is the same as neutrophils and other phagocytic cells found elsewhere in our bodies.

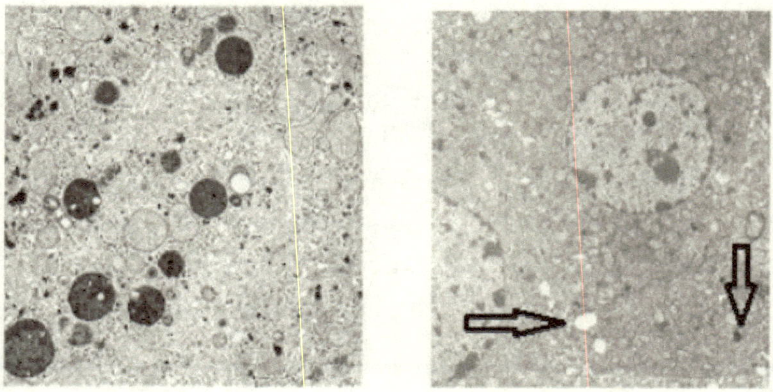

Sick liver cell full of lysosomes

Normal liver cell has fewer lysosomes

(Author's photos)

Figure (15) lysosomes in normal and in diseased liver cells

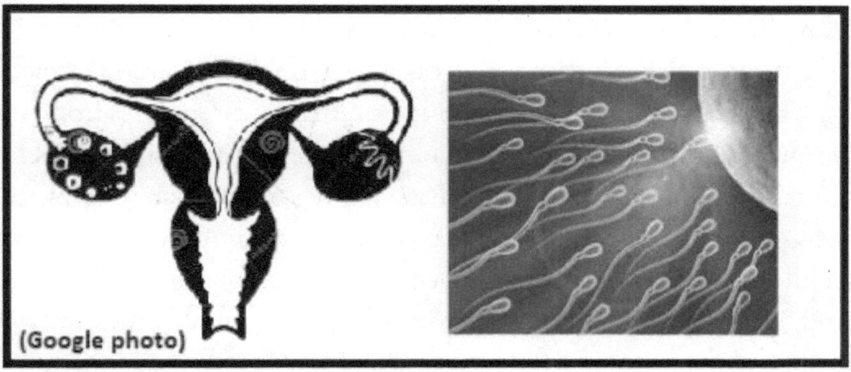

Figure (16) Female genital tract, sperms and ovum

Our next example is sperm. This tiny creature is in fact half a cell, not a complete cell and cannot be seen with the naked eye. Yet it goes through a strenuous journey in complete darkness eagerly searching for an egg (ovum) to penetrate and fertilize leading to conception. Recent research has confirmed that it is the ovum that chooses the sperm it 'feels comfortable with", not necessarily the first one to get there as is commonly believed. This led one scientist to describe this moment as 'another wedding' inside the womb.

The place inside the womb is dark, as we mentioned, and the path is too long relative to the sperm size. It is true that the scent of the progesterone hormone emanating from the ovum attracts and guides the sperm towards the ovum with some uterine contractions from the beginning, But the entire process is calculated and meticulously planned. It seems that both the sperm and the fertilized ovum have something like "living computerized mini brains" that enable them to fulfill their heavy tasks.

After Fertilization, the fertilized ovum becomes ready and committed to deliver a full baby after nine months according to a coded program of her own. And, since sexual reproduction results in infinite possibilities of genetic variation, each human couple, for example, has the potential to produce infinite number of children who differ from both parents and from each other If the couple age permits.

The filaria parasite: Filariasis is transmitted by mosquitoes from one person to another. The adult worm lives in the human lymph nodes for several years, during which it produces millions of tiny filarial larvae that swim in the blood and are easily taken by mosquitoes when they bite an infected person, transferring them in the next bite to a healthy one. And, since the adult worms live within the lymph nodes, they damage them, and the lymphatic fluid is hampered from returning to the heart, leading to swelling of limbs and tissues. Surprisingly, the young filarial larvae become more active in the blood circulation at night, coinciding with the mosquitoes' visit to humans to ensure the highest chance of transmitting the infection! This is yet another miracle that is hard to believe.

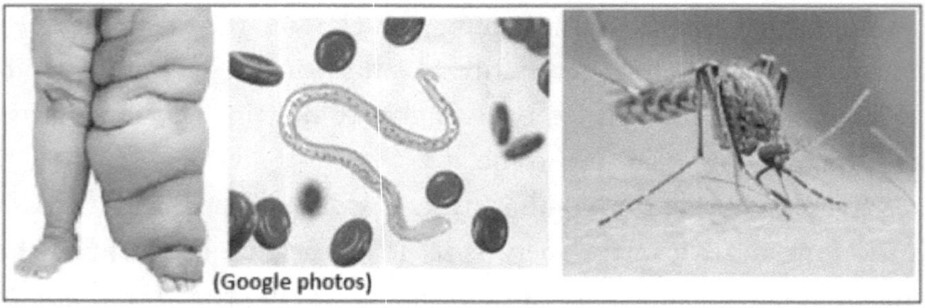
(Google photos)

Figure (17) Filaria transmission by mosquito

Now we shift to the world of insects. One species of the antlion fly lays eggs in soft soil like sand and dirt until hatching and releasing small larvae each at the size of a wheat grain. Each larva digs a neat conical pit for itself in sand and stays at its bottom waiting for an ant or small insect to fall in and eat it. The larva remains in that location for weeks without water, hiding in scorching summer sands without drying or death. It has been said that it must have a "miniature computerized air conditioning system"! After feeding and increasing in size, it transforms into adult fly that leaves the pit to repeat life cycle elsewhere.

The question here: How the fly that laid her eggs in the sand and flew away was assured about the future of her eggs and what would happen to them? And how did the larvae that emerged from the eggs knew what to do: to build a deep, smooth conical pit with sloping walls so that any insect that falls in cannot escape, thus ensuring her food? All those events are programed and planned in the fly's DNA.

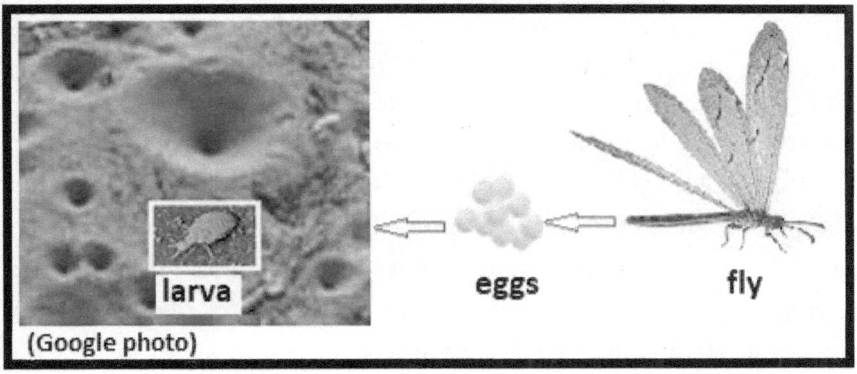

Figure (18) The life cycle of the Antlion fly

The enduring cockroach: Some scientists conducted research on brain and antennae of cockroaches. They found that antennae transmit six sensations simultaneously: smell, taste, temperature and humidity variations, touch, and gauging distance. The cockroach sneaks through dark sewer channels and uses its antennae like a blind person's cane. The cockroach's brain and body are adapted to the harsh conditions of its environment, not affected by darkness, infections or diseases, no matter the type of filth in their habitat. Furthermore, scientists discovered that the cockroach's body is even resistant to atomic radiation and concluded that it is the only creature that would survive full-scale nuclear war.

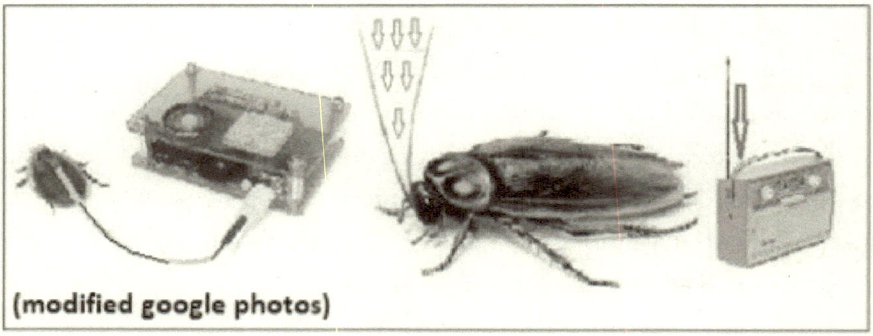

Figure (19) Research on cockroach brain and antennae

Since we are discussing miracles orchestrated by the DNA molecule, viruses must be brought here to attention. A virus is not a complete cell but a small piece of DNA or RNA enveloped by proteins. Scientists even disagree on whether to classify them as living organisms or not. Yet, they can kill millions of humans and paralyze the entire world as it happened during the COVID-19 crisis. Viruses exist on earth for two purposes: to spread, and to disrupt the DNA of others.

Finally, we should also have a brief note about plants. They are well known for their economic skill: enduring hunger and drought, seizing opportunities, so when food is abundant, they store the surplus, also play a significant role in providing food and oxygen essential for the survival of the planet. A scientist once said that plants can teach humans patience, perseverance, hope, and the art of seizing opportunities. Interestingly, some plants have been found to eat frogs, mice, and insects, hunting them with carefully devised strategies in the absence of a nervous system or "brain".

In a research article published earlier this year by Fredrico Cappini (https://blog.rapusia.org/animals/2817/why-do-plants-feel-pain-but-cannot-scream), he concluded that plants are capable of sensing damage, vibration, and have memories, also have underground communication, as well as air communication using volatile chemicals.

In conclusion, there is no secret, no magic, and every action or move or miracle linked to any living organism is attributed to its DNA molecule.

Figure (20) Plants relax with touch and plants trap insects

Summary

It is time to clarify nouns and terms in this aspect: The noun "AlRouh" is a collective term for three different entities: spirit, soul, and secret of life, each one is different from the other.

1- THE SPIRIT: In our view, the nature of spirit is not exactly known, but medically speaking, it is a "command center" in brainstem strongly linked to the moment of death, and it is a DNA product. When it stops controlling the body systems, death occur. It is possible though that mysterious energy emitted by DNA remnants left by the dead in areas where they lived could spark supernatural activities felt by the living. Whether there is weight loss or not at the moment of death is not scientifically confirmed.

Doctors all over the world don't know what comes out from the body at the moments of death, but they know exactly what happens at this moment: organ failure followed by brainstem failure.

2- THE SOUL: As the DNA material is found in every single cell in the body, it dictates all body activities through series of transcriptions and translations, along with organ specialization. The soul (or self) is built up from such DNA activities in addition to learning from the environment. The brain itself, which controls all body activities, is a DNA product.

3- THE SECRET OF LIFE: After death, a long series of activities continue in the cadaver. At first, all tissues and organs remain vital for a while and can be transferred and transplanted with confidence because they still have a strong "secret of life" with intact cell membranes and minimal cell loss. This secret fades rapidly unless tissues are frozen. The secret of life helps to bond transplanted organs into place and move with them from the dead to the living, it also ensures the continuation of the same function in the new place, regardless of where the DNA came from.

- The DNA molecule has a huge potential that does not go away by the death of its owner, but its influence and use continue in many ways.

For example, the continuation of life in transplanted organs, cloning, and as an accurate tool for identification, in addition to supposed supernatural powers it emanates.

Almost all countries have authorized organ donation and transplantation from the dead or the living under certain rules regulated by each society. For example, the Egyptian advisory opinion "Fatwa" No. 16177 in 2003 allowed organ donation and transplantation in Egypt.

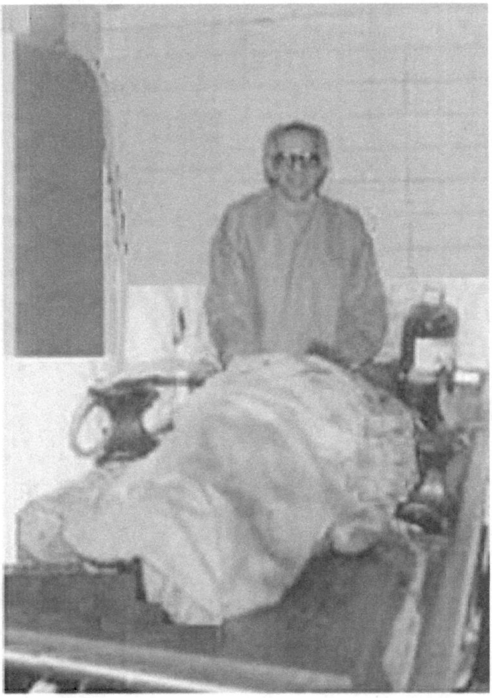

Picture of the author during one of his anatomy classes

References

1- Gray's Anatomy, 42nd ed, 2020 - Elsevier

2- Langman's Medical Embryology, 14th ed, 2018 – Walter Kluwe

3- “what is new about spirit and soul” – Tanta University Journal 2022 https://mkmgt.journals.ekb.eg[1]

4- Websites, DNA-Wikipedia

5- "الروح والطب الحديث"، الدكتور السباعي حماد - طبع وتوزيع دار المعارف 2015

6- "الروح"، ابن القيم الجوزية - مؤسسة المختار للنشر - 2006

Keywords: Spirit, Soul, Hammad, Theory, Secret of Life

1. https://mkmgt.journals.ekb.eg/